SIMPLE SUMS

Carol Watson
Illustrated by David Higham
Consultant: Wyn Brooks

Deputy-Head Teacher of The Coombes School,
Arborfield, Berkshire; lectures widely on
Primary School Mathematics.

This is Mr Zappo's circus.

He has lots of jugglers, clowns and acrobats.
2

Can you count the clowns?

How many people are there in
the picture altogether?

Mr Zappo is counting his acrobats.

How many acrobats are on the bottom row?
4

How many are there
on the top row?
How many in
the middle?

How many acrobats does Mr Zappo
have altogether?

5

José and his jugglers practise every day.

How many blue balls can you count?

ow many red balls are there?

How many balls are there altogether?

Bonko, the lion tamer, has 4 lions on one side and 5 on the other side.

If you count them and put them together, there are 9 lions.
We call this adding.

8

ou can write about lions in
nath-code.
he symbol for add is   **+**
he symbol for makes is   **=**

4 lions add 5 lions makes 9 lions.

4   +   5   =   9

Zoran has 2 icecreams.
Berti has only 1.

$$2 + 1 = \boxed{3}$$

Can you see what numbers are missing below?

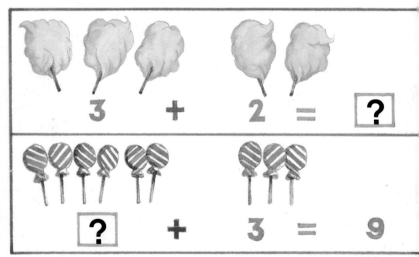

$$3 + 2 = \boxed{?}$$

$$\boxed{?} + 3 = 9$$

ow many clowns are in each group?
ow many are there altogether?

an you write this in math-code?

How many chimps are in each group?
Count 4 chimps. Count 1 chimp.
ut them together and count 5 chimps.

Wally is washing the elephants.

He has 4 elephants to look after.

One of them is clean. Some of them are still dirty. Count them.

How many still need washing?

There are 5 acrobats bouncing on the trampoline.

acrobats have fallen off.
How many are still bouncing?

Woppo the clown had 10 pies to throw at Willo.

He has thrown 3 pies.
How many are left?

Trigger, the horse, has eaten 2 more of Woppo's pies.

How many pies has Woppo left to throw?

Willo is selling balloons.

He had 8 balloons.
Joe has bought 2 balloons and is
taking them away.
18

balloons take away 2 balloons
akes 6. You can write this in
ath-code too.

he symbol for take away is −
he symbol for makes is =

8 − 2 = 6

4 lions are together. 2 run away. How many are left?

 − = ?

How many balls does the juggler have? He drops 3 balls. How many are left?

? − 3 = ?

What numbers are missing?

20

Bongo the chimpanzee has 6 bananas.
He is giving 2 to Coco.

How many bananas will Bongo have left?

Can you write this in math-code?

Woppo has
4 buckets
of water.

He drops
1 bucket.

How many
buckets
are left?

What is the math-code for this?

Can you do these simple sums?
Use the pictures to help you.

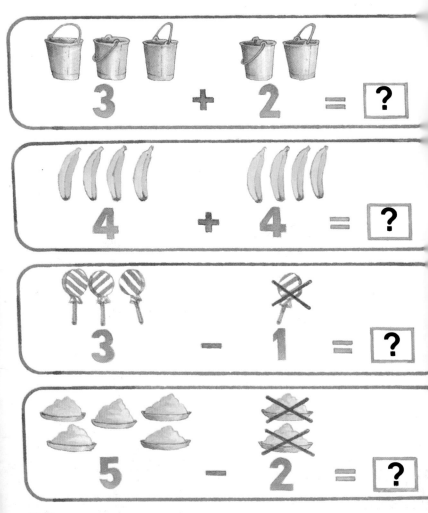

3 + 2 = ?

4 + 4 = ?

3 - 1 = ?

5 - 2 = ?

ow can you try these sums without
ctures?

5 + 2 = ?

7 + 1 = ?

2 + 4 = ?

5 - 3 = ?

7 - 1 = ?

6 - 4 = ?

ou can use beads or buttons to
lp you.

Simple sums
Can you think what numbers are missing from the boxes?

$3 + 2 = \boxed{?}$ $5 - 2 = \boxed{?}$

$2 + \boxed{?} = 4$ $4 - \boxed{?} = 2$

$5 + 3 = \boxed{?}$ $8 - 3 = \boxed{?}$

$\boxed{?} + 4 = 6$ $\boxed{?} - 2 = $

$7 + 2 = \boxed{?}$ $9 - 7 = \boxed{?}$

$6 + \boxed{?} = 7$ $7 - \boxed{?} = 1$

First published in 1984
Usborne Publishing Ltd
20 Garrick St, London
WC2 9BJ, England
© Usborne Publishing Ltd 1984

The name of Usborne and the
device 🎈 are Trade Marks of
Usborne Publishing Ltd.
Printed in Portugal

Lyke Wake Walk

THE OFFICIAL GUIDE OF
THE LYKE WAKE CLUB

Dalesman

Dalesman Publishing Company Ltd
Stable Courtyard, Broughton Hall,
Skipton, North Yorkshire BD23 3AZ
www.dalesman.co.uk

First published 1959
This updated edition published 2001

© Bill Cowley, 1959, 1993; Paul Sherwood 2001

Maps by Jeremy Ashcroft
Cover: Cringle Moor by Ken Paver

A British Library Cataloguing in Publication record
is available for this book

ISBN 1 85568 191 9

Printed by Amadeus Press, Cleckheaton

PUBLISHER'S NOTE
The information given in this book has been provided in good faith and is
intended only as a general guide. Whilst all reasonable efforts have been made
to ensure that details were correct at the time of publication, the author and
Dalesman Publishing Company Ltd cannot accept any responsibility for
inaccuracies. It is the responsibility of individuals undertaking outdoor activities
to approach the activity with caution and, especially if inexperienced, to do so
under appropriate supervision. They should also carry the appropriate
equipment and maps, be properly clothed and have adequate footwear. The sport
described in this book is strenuous and individuals should ensure that they are
suitably fit before embarking upon it.

CONTENTS

Origins of the Lyke Wake Walk 5

The Cleveland Lyke Wake Dirge 11

The Lyke Wake Club 13

Conditions for a Challenge Walk 21

Races and Records 24

The Route 27

The Shepherd's Round 40

The Hambleton Hobble 44

URRA MOOR

I walk alone through moor and bog,
Rain and wind sweep all around,
Walking through clouds that touch the ground,
Heather high, heather low,
Heather round my feet below,
A bleating sheep, a screeching grouse,
The fleeing of a small field mouse,
A hundred times I've trod this moor,
Along the tracks of yesterday,
Then is it right for me to say,
No man knows all the way.

JOHN E. GRAY

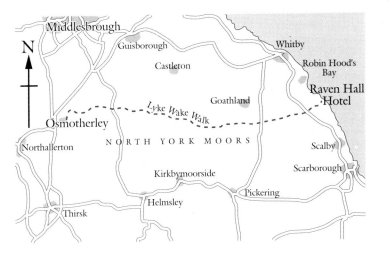

ORIGINS OF THE LYKE WAKE WALK

Bill Cowley

For me, as for anyone born on Teesside or on the Cleveland Plain, the Cleveland Hills were the backcloth of boyhood. The nearest I ever got to the present Lyke Wake Walk was a three day walk from Danby to Troutsdale and back through Farndale to Broughton, in the late 1920s.

Those who would enjoy the Lyke Wake Walk, and other such walks, to the full, must know the country intimately: its geography, its geology and its history. The North Yorkshire Moors contain 1,550 square kilometres of wild moorlands and lovely valleys, many of them quite uninhabited. A lifetime is insufficient to know the whole area as it should be known. Here ancient superstitions, and an ancient language, lingered till very recently.

From the long fertile Vale of Pickering rise first the Tabular Hills of coralline limestone, whose steep northern scarps, from Black Hambleton to Barns Cliff and Silpho, are always prominent in the Lyke Wake landscape to the south. North of the Tabular escarpment rolls a sea of heather-clad moor, one of the largest uncultivated tracts in England, the dominant feature of the district from time immemorial, and the main feature of the North York Moors National Park.

Eight hundred years ago it was known as Blackamore. John of Hexham, writing in the days of Henry II, states that Rievaulx Abbey was situated "in solitudine Blackaumor". Drayton's Polyolbion in 1622 mentioned "large-spread Blackimore".

Geologically, the Tabular Hills are Middle Oolite and their conspicuous character is due to hard beds of Calcareous Grit and Coralline Oolite on top of 50 metres of Oxford Clay, itself above a band of Kelloway Rock. The high moors are Lower Oolite, Moor Grit and Fossiliferous Grit on top of thick beds of yellow Estuarine Sandstones, separated from the underlying Lias by that very variable band, the Dogger. Though the main Cleveland iron seams are in the Lias, the Rosedale seams were in the Oolite – in the Dogger – and there are also occasional small pockets of coal and limestone.

The Lias beds are the foundation of our moors and give them much of their character: the steep northern scarp of the Cleveland Hills, and the deep narrow valleys. At the top of the Lias, just below the ferruginous band of Dogger, occur the alum shales, thirty metres thick. In the 16th century the Chaloners of Guisborough brought home to Cleveland the method of alum extraction.

New and cheaper processes killed this industry in the middle of the 19th century, but there are many remains of alum and jet workings to be found throughout the area.

Below the alum shales are the Jet rocks, a series of shales sometimes so bituminous that they ignite spontaneously, as on Boulby Cliff. Hence the red colour of some shale tips. Jet itself is of uncertain occurrence, and though plenty of jet ornaments are on sale in Whitby they are mostly of Spanish jet. Yet in 1872 jet working employed 1,500 persons and the value of the trade was £88,000. There is a Jeater Houses between Thirsk and Osmotherley, and a Jet Miners' Inn at Great Broughton. 'Striking a good seam' had a particular local application in Cleveland speech.

Probably the last to be found and commercially worked was by an aged jet miner in Rosedale, about 1885, near the Hamer road. Having no legal rights, he worked it at night, and succeeded in reaping the benefit of his experience and observation. Apart from iron, jet is the characteristic mineral of our moors. Nowhere else in England does it occur in such quantity or quality, and it has been known and used since Neolithic days. Prehistoric man, ancient Briton, Roman, Saxon, Dane, Viking and Norman have all admired and worn Whitby jet.

The high moors crossed on the Lyke Wake Walk are of enormous antiquity, having been a land surface since the close of the Chalk Period, three to six million years ago. Tropical plants grew on them in the early stages and the petrified trunk of a large cycad tree was exposed in a sandstone quarry below Scarth Wood.

There are one or two traces of Stone Age men just before or during lulls in the Great Ice Age, and the last survivors of these ancient hunters left pygmy-flints in moorland camps as at Cockheads, near Hamer, round 6000BC. Like stray shafts of sunshine, these small imperishable flints illuminate the depths of gloomy avenues of time.

Early Bronze Age man followed them onto the limestone hills and elsewhere, and later the long-barrow men, but they never inhabited the high moors. It was not until about 1000BC that the mid-Bronze Age round-barrow people, who had come over from Denmark and the Rhinelands, began to inhabit the moors, driven to those inhospitable heights by other invaders from Central Europe, the Brigantes.

About 400BC the Parisii came from France and occupied the Wolds, driving the ancient urn people who remained there onto the moors. When the Romans came to conquer the Brigantes and the Parisii, descendants of the urn people were still clinging to the high moors, living on in the primitive fashion of their ancestors. The Romans left the high moors untouched save for Wade's Causeway.

With the coming of the Anglo-Saxons some Britons in their turn had to flee to the moors, which likewise must have long remained a British island in a Saxon sea, as many Celtic place names bear witness.

There was a new series of Scandinavian invasions, with the Danes in the ninth century, and then the Norwegians. Some came direct, and some came from Ireland, through Cumberland, bringing with them Gaelic-Scandinavian place names. Farndale may be one. Barnscliffe, the cleft through which the waters of Bloody Beck, Jugger Howe Beck and the River Derwent find their way; the Gaelic bearna, a cleft between two hills.

It is probable that Norwegians repopulated parts of North Yorkshire after its harrying and devastation by William the Conqueror, and that immigrants were still coming in the 12th century. Except for a few British names, the Scandinavian influence is paramount in our place names as in our dialect. The few Britons who remained have long since vanished, probably not exterminated but just swamped by the Norwegians, who were attracted to the moors and the dales perhaps because of the resemblance to their native land.

The Britons left a few names and some traces of Celtic fields near Commondale, Wardle Rigg, and Cloughton. But the Bronze Age round-barrow people or urn folk have left their monuments all over the moors. Their barrows or howes (O.N. haugr, a mound), and their standing stones, are our landmarks from one end of the Lyke Wake Walk to the other, though most have been given Norse names by later invaders. These people buried some of their dead unburned, the ordinary folk perhaps, but the chiefs and leading members of the tribe were all cremated before burial, and their remains placed in urns under the barrows which were often sited on territorial boundaries (as they still are in the form of estate boundaries).

There could be up to 10,000 barrows of various sizes on these uplands. The idea of cremation was probably to prevent the ghosts of the dead from returning to trouble the living, and some of this superstition lingered on in the dales until modern times in the throwing of charcoal into graves. It is perhaps echoed in the Lyke Wake Dirge –

Fire an' fleet an' cannle leet'.

Sometimes these burials are also marked by stone circles like the one on the ridge between Bilsdale and Tripsdale, the Bridestones. Other stone circles and standing stones are connected with ceremonial – perhaps sacrificial – rites. Whether always associated with fertility rites or not, many of the standing stones must have been useful then, as they are to us now, as landmarks arfd guideposts, for no doubt the sea-roke (mist) often came down on Botton Head and Shunner Howe as it does today!

Many of the howes and stones lie in long lines, as on the ridge south from Carlton Bank, and between Ryedale and Bilsdale, or right across Fylingdales from the Bridestones above Grosmont to the barrow groups on Maw Rigg above Langdale. Still more remarkable, however, are certain well-marked settlement sites on Danby Rigg, Crown End above Baysdale, and in many other places where whole cemeteries of barrows, entrenchments, hollow cattle-tracks, hut-sites and the remains of small fields have been traced. One of these is at the beginning of the Lyke Wake Walk, near Scarth Nick, and one at the end, between Pye Rigg Howe and Peak.

Some of the urn folk's sites are those of farms inhabited to this day. They themselves had sheep and cattle and a primitive agriculture. Their pathways ran mostly along the ridges, north and south, but there is little doubt that they would use the high watershed for east-west crossing, and that we tread often in their footsteps. The Cleveland historian, the late Frank Elgee, says, "Their ultimate fate is shrouded in darkness. Somewhere, at some time, a woman's deft fingers moulded the last urn, the wind drifted the smoke of the last funeral pyre over the moors, and the last cairn was piled over the dead. "We who walk the high moors now cannot but be conscious everywhere of these long vanished people."

On returning from India in 1947 I bought a 21 acre smallholding at Over Silton, below Black Hambleton, and wrote in *The Dalesman*, that from the moor gate at the top of the village a man could walk for three days through heather, to the sea near Whitby, and never meet a soul. Later I had an interest in Glaisdale Head Farm. One evening in 1955 I climbed up alone to the rigg between Glaisdale and Fryupdale and whilst digging peat on Cockheads I thought of the forty miles of heather between my main farm, at Swainby, and the sea.

The moors between undulated endlessly, dotted with tumuli and, as always at such time, I began to imagine them peopled with ghosts of a remote past. What rites and ceremonies had marked the burial which had taken place so long ago on the mound where now I stood.

> Grey recumbent tombs of the dead in desert places,
> Standing stones on the vacant wine-red moor,
> Hills of sheep, and the homes of the silent vanquished races,
> And winds austere and pure.

The Lyke Wake Walk began with a further article in *The Dalesman* for August 1955 and the challenge to walk this in twenty-four hours was soon taken up; the first crossing occurred later that year.

At noon on October 1st, ten men and three women set out with me to try to do the complete traverse of the moors in 24 hours. We knew it would be a tough test – a minor Everest of our own making. We had all lived with maps in hand or mind for days, weighing alternative routes. Now came the final choice – and the weather was perfect.

One fine but difficult route, over Whorlton and Snilesworth Moors to Chop Yat, was ruled out by a shooting party, so we plunged down to Scarth Nick and into a sea of bracken.

We took the alum miner's track round the face of Carlton Moor, Cringle Moor and Cold Moor, with the Cleveland Plain below stretching away to the Pennines by Cross Fell, or across smoky Teesside to the Durham hills. Four of us reached Clay Bank Top in 2$^{1}/_{2}$ hours, glad of the mobile bar that awaited us!

We reached the high point of Botton Head (1,489 ft.) right on schedule at 3.30 p.m. Now we were deep in the moors and deep in heather. All our concentration was required to find the easiest and the shortest way through, a Bronze Age mound or a leaning stone our guide, to the Smugglers' Trod, its stone flags now heaved crazily about by heather roots. Darkness was just closing in on us as we saw the lights and tents of our bivouac round the ruined inn of Hamer. We gave a great shout and rushed down through the heather. We had covered 21 miles in seven hours and had earned a rest. The Cleveland Lyke Wake Dirge came to our minds.

Few of us got much sleep. The worst part still lay ahead. The real testing time of this walk started at 3.30 a.m. Before us, as we struggled through deep heather with storm threatening, was the wild stretch of Wheeldale Moor with never a track across it – just four miles of knee-deep heather till we reached the Roman highway at the other side.

Clouds veiled the moon. The light and the contours were deceitful. We felt rather than saw the sudden drop into Wheeldale Gill, our guide to the left. To the right was only the cold night wind on our cheeks. Startled grouse kept exploding from the heather at our feet and grumbling away into the darkness. Shallow dips seemed like deep valleys, and sometimes we fell full length in an unseen hole. We kept checking our course by compass, slightly south of east.

One solitary light in Goathland gave us another check, and then suddenly an intake wall loomed ahead by the Stape road.

Beyond was Wade's Causeway and the steep track down to a dark and silent Wheeldale Lodge. At 5.30 a.m. we were sitting on the stepping stones chewing chocolate, the wide stream a subdued silver between rustling trees. There was another hard climb through rocks and bracken onto Howl Moor. Dawn broke slowly as we approached the railway cutting at Ellerbeck. We crossed by Fen House, over Tom Cross Rigg to the Whitby road. A rest and a cold breakfast, and we set off along the Salt track for Lilla Howe.

It was 10 a.m. as we crossed the Scarborough road near Helwath Bridge and knew we were well inside the time. Even so, the last rough patches of heather over Pye Rigg Howe were a trial, and it was with weary limbs and sore heels, but glad hearts, that we tramped into Ravenscar at last, down the road from the old windmill.

Blue sea and golden sands were just beginning to show through the mist as the sun gained strength, and the cliffs beyond the bay took shape as we celebrated our triumph with the Goulton Tankard and formed the Lyke Wake Club.

But what of the name?

THE CLEVELAND LYKE WAKE DIRGE

This mediaeval dirge was mentioned in 1686 by John Aubrey as being sung at funerals "by the vulgar people in Yorkshire" although it had almost ceased to be sung or chanted at funerals in the area by the mid 1600s. It is thought that it was last sung at a funeral in Kildale in about 1800. However, it was sung in a howling gale on Cringle Moor in October 1985 as Club members scattered the ashes of a well-respected senior member of the Lyke Wake Club.

Several versions of the dirge are known throughout northern England.

The ideas in the dirge go back to very early folklore – perhaps even to the Bronze Age people who burned and buried their dead on the high points of our moors, where their grave mounds or 'howes' are now our guiding marks. Apart from these burials, there is no suggestion that corpses were ever carried over the length of the Lyke Wake Walk.

Appropriately it suggests that everyone, after death, has to make a journey over a wide and difficult moor. If you've done good deeds during your life – given away socks and shoes, silver and gold, food and drink – aid will be given you and you'll get across safely, to Paradise; but if not, then you'll sink into Hell flames.

> This yah neet, this yah neet,
>   Ivvery neet an all,
> Fire an fleet an cannle leet,
>   An Christ tak up thy saul.
>
> When thou fra hence away art passed,
>   Ivvery neet an all,
> Ti Whinny Moor thoo cums at last,
>   An Christ tak up thy saul.
>
> If ivver thoo gav owther hosen or shoon,
>   Ivvery neet an all,
> Clap thee doon, an put em on,
>   An Christ tak up thy saul.

Bud if hosen an shoon thou nivver gav neean,
  Ivvery neet an all,
T'whinnies'll prick thee sair ti t'beean,
  An Christ tak up thy saul.

Fra Whinny Moor when thoo art passed,
  Ivvery neet an all,
Ti t'Brig o Dreead thou cums at last,
  An Christ tak up thy saul.

If ivver thoo gav o thy siller an gowd,
  Ivvery neet an all,
On t'Brig o Dreead thoo'll finnd foothod,
  An Christ tak up thy saul.

Bud if siller an gowd thoo nivver gav neean,
  Ivvery neet an all,
Thoo'll doon, doon tummle towards Hell fleeam,
  An Christ tak up thy saul.

Fra t'Brig o Dreead when thoo art passed,
  Ivvery neet an all,
Ti t'fleeams o Hell thou cums at last,
  An Christ tak up thy saul.

It ivver thoo gav owther bite or sup,
  Ivvery neet an all,
T'fleeams'll nivver catch thee up,
  An Christ tak up thy saul.

Bud if bite or sup thoo niver gave neean,
  Ivvery neet an all,
T'fleeams'll bon thee sair ti' t'beean,
  An Christ tak up thy saul.

Canon Atkinson of Danby believed that 'fleet' (in "Fire an fleet an cannle leet") was a variant of 'flet', a Cleveland term for live coals or embers, some of which in his early days in Danby (1840s) would still be thrown into the open grave.

'Wake' is the watching over a corpse, and 'lyke' (*leich* – Old Germanic) is the corpse itself – as in the 'lych' gate of a church where the coffin was rested.

THE LYKE WAKE CLUB

The Lyke Wake Club must be unique in that it has no formal organisation and no subscription. It has been likened to a tribal society – entry into which is by ordeal!

There is perhaps no other walk which covers such a complete and well-defined stretch of hill country within the compass of one long day. All who complete the walk in the conditions laid down become members, or Dirgers & Witches – currently about 160,000; there are no honorary ones.

People sometimes ask why the Club doesn't keep a tighter time check on people doing the walk. For one thing, of course, it's impossible; however, it is quite unnecessary. No one who attempts a walk of this kind is going to cheat. If they did, they would not last half an hour at a Wake amongst people who know every inch of the way, on every alternative route.

The challenge and the satisfaction are your own. It is a shared experience of the same walk on differing routes and in differing conditions.

That gives a special basis of fellowship to the Lyke Wake Club.

The Club confines its activities to collecting information about the Lyke Wake Walk, the Shepherd's Round and the Hambleton Hobble. It is also concerned, however, with encouraging members to learn all they can about the North York Moors, its history and folklore, and to assist in safeguarding this area.

The Club has gradually built up, with more humour than seriousness, its own rather macabre traditions based on the Cleveland Lyke Wake Dirge and other local folklore. The crest is a silver coffin and three silver tumuli on a black shield; the badge is a silver candle and two tumuli on a black coffin.

Women members form the Circle of Witches whose duty is to cast suitable spells and ward off the machinations of Hobs, Boggets, Gabriel Ratchets and the like.

In explanation of some of this folklore see the story of the Glaisdale and Farndale Hobs in Atkinson's *Forty Years in a Moorland Parish* published in 1891. There, too, you will find much about the witches of this district who so frequently turned themselves into hares.

The emblem of the Club is the rowan – mountain ash or witchen tree.

The Club Mace, presented in 1960 by Selby Round Table, is of polished rowan with a facsimile of the Scarth Wood Moor Ordnance Survey pillar on one end, and a coffin on the other. It has been carried across the route on at least two occasions!

From the early days of the Club a few enthusiasts used to gather at the *Queen Catherine* in Osmotherley for what might be called a post-mortem of the year's efforts. This became the annual Winter Wake, now the Candlemas Wake in early February; the business meeting and annual dinner.

Over the years there have been Midsummer Wakes at several locations and many anniversary Wakes in October, at such places as Hull, Scarborough, Rosedale, Ravenscar and Northallerton.

For many years an annual Brewster Sessions Wake was held in a Sheffield Brewery, and some members have had 'licensed' Wakes in their own areas, which have occasionally been graced by Club Officials.

The correct dress for Wakes is of course black, but in default of this some suitable sign of mourning should be worn, and it should never be forgotten that the Wake is a solemn occasion. Any display of mirth is most unseemly, and sympathy should be indicated by deep and heartfelt groans.

Prior to the meal of jugged hare, wild boar or less ambitious fare for lesser mortals, the Cleveland Lyke Wake Dirge is sung. After the meal is The Loyal Toast, given by the Most Miserable and Melancholy Mace Bearer to "Edward II and the Lady of the Manor of Goathland", the relevance of the toast being that Edward II had hunting forays on these moors in 1323 and through the Duchy of Lancaster; the Queen is the Lady of the Manor of Goathland!

The Wretched and Erroneous Recorder disinters the 'Centuries' of the previous Wake, and a post-mortem of the year's events takes place followed by the conferring of degrees on dirgers and witches.

So many people have done it a great many times that the Club has instituted an amusing – but none the less arduous – series of degrees. To become a Master or Mistress of Misery (black neckbands) you must do three crossings, one of which must be in the opposite direction. To become a Doctor of Dolefulness (purple neckbands) you must do four more crossings, one of which must be in winter, and one unsupported. A further degree is that of Past Master, for which you need fifteen crossings and have contributed great services to the Club.

A Past Master should be able to find his way across any moor by day or night, without map or compass, drunk or sober!

Four people have carried on to become Senile Centenarians. Louis Kulscar was the first to complete one hundred crossings in November 1973 – at least three of his crossings were done barefoot! A year later Ben Hingston reached his century. These were followed shortly after by Ian Cooper and Gerald Orchard.

In January 1983 Ben Hingston completed his 200th crossing and became the first and only Bi-Centenarian. He completed his 212th, sadly his last, on 9th February 1984.

The Lyke Wake Club has now been in existence for almost fifty years. The 'Centuries' of over eighty Wakes occupy a vast tome. Short extracts from four give an indication of the activities and historic events over the past four decades.

SELECTED CENTURIES OF ANCIENT WAKES

Of the Eighth Wake (being the Second Midsummer Wake) held at Potto Hill, 23rd June 1962.

The Reception was held on the occasion of this Wake in a Labyrinth of Barns located in the grounds of the Chapel of Rest of the Chief Dirger. The precincts had been laid out in a truly funereal fashion, with a ticket collector reposing in a coffin at the entrance. The Barns were complete with antechamber for Masters of Misery and Minstrels Gallery where voices were raised in Bacchanalian Song until well after midnight.

The Dinner was presented in fine style, buffet fashion, within the confines of the Barns. It consisted of the usual choice fare now associated with these funereal gatherings on Midsummer Eve. Cringle Moor Crisps followed Cock Howe Cockerels on the Bone, but preceded Grilled Green-howe Sausages. In order of gastronomity there then followed Sil Howe Salad, Blue Men i't Moss, Simon Howe Surprise and Bloody Beck Disaster.

The Address by the Chief Dirger commenced promptly and most unbelievably, at 9 o'clock, and lasted even more unbelievably for only half an hour. He proposed the usual toast to Edward II and the Lady of the Manor of Goathland, and then called upon the Wretched and Erroneous Recorder to intone the Centuries of the last Wake.

Unfortunately, that Miserable individual had been prevented from attending, but he had sent his apology for absence and also made adequate provision for all his duties to be undertaken by a Deputy Sorrowful Recorder who, at the call of His Mournfulness, discharged his duty with commend-able alacrity and somehow succeeded in convincing even the Chief Dirger that what he was saying was true and correct.

The Chief Dirger next placed on record some of the more noteworthy reports of crossings received, indeed, the rafters tolled with misery as His Mournfulness recounted how 120 new recruits in the Army had succumbed one by one until ultimately only three finished. Another Army party had done a crossing and reported sighting Gamekeeper Wilson sitting on a Howe glowering at them, but he neither moved nor said anything. Another party had described the crossing as 'the finest bog trek in the North'. The Chief

Dirger finally reported that he had himself taken across the first party of farmers.

The Address by the Sorrowful Shroud Supplier was given at 9.40pm. He explained that the stars on the tie were correct depicting five and six point tumuli, but that those on the badges, owing to the manufacturer's error, were incorrect.

The Address by the Anxious Almoner was as short as the Funds are low. He affirmed that they continue in their customary sorrowful state.

His Mournfulness first disposed of his grievous task of bestowing the degrees of Master of Misery and Doctor of Dolefulness. His own son John was admitted a Master of Misery despite the fact that he had only attended one Wake, although he had done three crossings. It was considered that as he had drunk cider at the age of three, and beer at the age of seven, he was adequately qualified.

Kenneth Smart from Hull then ascended the Rafters and from his funereal eyrie gave his Doctor's Dissertation, reading from what it was thought was going to be an unending roll of toilet tissue, pausing only at the perforations.

(Ted Jefferson)

Of the Thirty-second Wake (being the 18th Winter Wake) held at the Queen Catherine Hotel, Osmotherley, in the North Riding of the County of York on the 8th day of December in the year 1973: Incorporating also the non-existent Decades of the Michaelmas Wake held at Potto Hill, Swainby, in the aforesaid Riding on the 29th day of September of like year.

The Assembly: Dirgers and Witches foregathered according to their wont to imbibe the brews and distillates according to their taste and, hopefully, within their capacity.

The Banquet: Jugged Hare chased Witches Broth and was pursued by Crab Apple Pie, etc., in time-honoured ritual.

The Toast: His Mournfulness, the Chief Dirger, proposed the customary Loyal Toast to King Edward II of 1323 renown and the Lady of the Manor of Goathland. There followed a rendition of the Lyke Wake Dirge, led by the Dirgers in Chief, before the Chief Dirger launched forth on his address.

The Sermon: Reporting the retirement of, and paying tribute to the literary works of the Wretched and Erroneous Recorder, Ted Jefferson, His Mournfulness confessed that his recollection of the previous Wake, which had passed unrecorded, had faded in the mists of time and he enjoined the Harassed Archivists – or one of them – to disinter the Centuries of some recent Wake.

The Harassed Archivist, prompted dutifully by his spouse, the Harassed Archivistess, recited the Centuries of the Midsummer Wake, the thirtieth, held at Potto Hill on the 23rd day of June in the year 1973 and since none could gainsay their veracity nor indeed vouch for it these Centuries were duly accepted and appropriately endorsed by His Mournfulness.

His Mournfulness next turned to the subject of the Club accounts. In August, he reminded us, we had suffered the grievous loss of Lennard Douglas who had been Anxious Almoner since the inception of the post at the Winter Wake in the year 1961. Recalling that over the twelve years, Lennard had shared many a pint, the Chief Dirger proposed a toast to his memory and the assembled company rose to quaff in deferential response.

Reverting to the accounts, His Mournfulness recounted that in late August, we had found ourselves the luckless legatees of all Lennard's ledgers and stock – and a bank account in his name, automatically frozen. The accounts were titivated professionally by Derek Grass and it was his suggestion that a Lyke Wake Company should act as a management company for the Lyke Wake Club.

Fortunately, Peter Long, Dirger of long standing; veteran of many crossings; survivor of myriad wakes and expert in Russian which in some peculiar way might qualify him for the onerous task of dealing with the Club accounts, took up the mantle of Anxious Almoner and, at this point in the proceedings, rose to present the accounts. The Anxious Almoner hastened to allay anxiety by explaining that the Reserve Fund, into which had been poured the princely sum of £1,000, being the major portion of the club's wealth, had been set up to finance possible legal action on rights of way. With the interest accruing from this fund, the Lennard Douglas Memorial would be founded and, at the suggestion of His Mournfulness, a sub-committee, consisting of Peter Long as treasurer and Ian Ashley Cooper, was appointed, to which Paul Sherwood was subsequently co-opted. His Mournfulness gave his blessing to a 'natural' break.

On resumption the accounts were summarily accepted and His Mournfulness commenced upon his customary post-mortem. Record is, however, penned of the one hundredth crossing of one Louis Kulscar, accomplished on the 17th day of November in the year 1973, and marked by a small ceremony at the Queen Catherine Inn. Following further light-hearted and light-headed discussion it was agreed that an apposite degree to confer upon the venerable old master would be that of 'Senile Centenarian' and it was reported that the well-known octogenarians Ian Ashley Cooper and Ben Hingston were hard on his heels.

The gathering closed following the bestowing upon the retired Ken and Margaret Morgan of gifts as tokens of esteem and gratitude for their services during their incumbency of the Queen Catherine temple; the thirty-second Wake subsided into a state of expiration.

(John Scarsbrooke)

Of the sixty-first Wake being a Peculiar Anniversary Wake, held at the Crown Hotel, Scarborough, in the County of North Yorkshire, on the 5th day of October 1985.

A Wake to celebrate the first 30 years of Lyke Wake walking! Held in the opulent gloom of this Quality International Hotel of which there are 800 worldwide, the Wake was attended by the faithful funereal few, augmented by ethnic minorities from the far-flung outposts of His Mournfulness' empire – Huddersfield, Pontefract and Scarborough.

It was not attended by the expected John Wilford nor by other unexpected absentees and only partly so by the supported Gerald Orchard who had apparently been engaged in yet another crossing for Aids Relief – or Relief Aid – and had brought with him his own supply of Lucozade, doubtless for the same purpose. The meal was of typical Lyke Wake Fare:

Caribbean Cocktail
(the rear end feathers of a West Indian male)

Roast Chicken Grandmere
(Chickens Grandmother or Old Boiling Fowl)

Chef's Selection of Vegetables
(Picked freshly from the fridge)

The Malevolent Macebearer did, as is his custom, with his mace. His Mournfulness the Chief Dirger proposed the loyal toast to Edward II and, as is his custom, left the Lady of the Manor of Goathland untoasted, subsequently rectified with ridicule by the Malicious Macebearer.

The Pro-Vice Chancellor and Suffragan Recorder disinterred the Centuries of the Wakes of the preceding annus. The Recorder Re-incarnate proposed an irreverent toast to Piers Gaveston who, before losing his head, had organised the first Scarborough Wake 673 years earlier. Whilst the Anxious Almoner maintained a discreet silence.

Opening his necroscopy the Chief Dirger reported that one Jack Russell did an unusual double, in one direction, accompanied by his 3 veterinary surgeons that suggested a Lyke Wake dog collar to be awarded vicariously.

In the Lyke Wake Race 96 participants finished roughly in the order pre-ordained, but most sadly George Ward, the much-respected Carlton blacksmith, succumbed to a heart attack on the railway becoming only the third in thirty years to make the ultimate crossing whilst on the walk.

The Melancholy Macebearer brought grave news of the failing health of the second Senile Centenarian and only Bi-Centenarian, Ben Hingston, whom he had recently visited in Northallerton Hospital. Ben, he reported, wished to subsidise a light-hearted and good-natured Wake to be held after he had given up the ghost, but was reluctant to send the cheque to the Chief Dirger since he didn't wish to burden the old man with more problems.

Recalling that Ben had entered every race since 1967 until 1984 when he became too ill, the Mace-toting Lyke Wake Race Secretary expressed his intention of inaugurating a Ben Hingston Memorial Trophy to be presented every year to the winning veteran.

The hour being late, the end of the Wake was marked by the intoning of the Dirge and whilst the rudely healthy disported themselves in the bracing East Coast air of the Scarborough beaches, the more adventurous repaired to the cosmopolitan clag of the hotel bar – though not, it should be noted, at any expense to the path repair fund.

(John Scarsbrooke)

Of the seventy-sixth Wake being the fortieth Winter Wake and the Bill Cowley Memorial Candlemas Wake, held at the Golden Lion Hotel, Northallerton, on Saturday February 4th 1995. Not since the halcyon days of the Potto Hill and Queen Catherine Wakes had so many dirgers gathered as came this day to pay homage and celebrate the life of the Chief Dirger, who began his journey over Whinny Moor on August 14th 1994.

The Golden Lion adaptation of the traditional menu proved bounteous and iniquitous providing sustenance for the extended memorial proceedings to follow. The apprentice Horn Blower, John Severs, and the Mace preceded the loyal toast to Edward II and the Lady of the Manor of Goathland, which in turn was followed by a seriously depressing dirge. The Centuries of the 39th Winter Wake were disinterred and were deemed to be sufficiently factless.

The Memories; The Anxious Almoner, Paul Sherwood, took precedence on the floor, stating that he and the late Chief Dirger had been friends for 35 years, and he reflected on Bill's final few weeks. Speaking of Bill's time in India before 1947, Paul had visited 'Tara Devi' where Bill started a scout camp, and had found many photos depicting the great 'Sahib Cowley'.

Looking to the future, he agreed with the many comments he had received that there should be no new Chief Dirger: Bill could not, and should not, be replaced.

In conclusion, before passing on to John Wilford, Paul played a recording of Bill reading his own evocative dialect poem *Epitaph to a Countryman*.

John Wilford the eloquent rose giving condolences to the Cowley family; John recalled that he had first met Bill when he produced the Lyke Wake film for Yorkshire Television, during which Bill had affectionately been known as 'Owld Grump'. Paul Cowley – Bill's grandson, who received a remarkable portrait of Bill by Ian Watters, a misguided founder member – made response to these tributes.

The seriously protracted proceedings were terminated by a dreadfully dreary dirge led by John Scarsbrooke.

(Mike Parker)

Shortly after the death of Bill Cowley two memorials were installed; Lord Ingleby of Snilesworth placed a carved stone on the Lyke Wake track above Scugdale, and the Lyke Wake Club donated a timber seat to the Great Yorkshire Showground at Harrogate. In October 2000 Lord Ingleby had a three-metre stone carved by the Weatherill family and erected on Snileworth Moor: 'Bill Cowley, A History of Snilesworth 1993' in recognition of his historic research. One face carried the ADMM millennium marking.

The present hierarchy of the Club is constituted as follows:

> The Misguided Foundation Members
>
> The Anxious Almoner (Paul Sherwood)
>
> The Horrible Hornblower (John Cowley)
>
> The Most Miserable and Melancholy Mace Bearer (John Scarsbrooke)
>
> The Wretched and Erroneous Recorder (Mike Parker)
>
> The Pro-Vice Chancellor & Chairman of the Court of Past Masters (John Vine)
>
> The Miserable Bier Carriers (a handful of senior members)
>
> The Harassed Archivist
>
> The Melifluous Minstrel
>
> The Senile Centenarians
>
> Past Masters in order of Passing
>
> Doctors of Dolefulness in order of Dole
>
> Masters of Misery in order of Mastery
>
> Witches in order of Witchery
>
> Dirgers in order of Tribulation

The Lyke Wake Company Limited was incorporated in June 1973 to safeguard the property rights of the Club.

CONDITIONS FOR
A CHALLENGE WALK

The Challenge Walk is the complete traverse of the North York Moors along the main east to west watershed, within 24 hours, although 24 hours of daylight are acceptable for a ski crossing.

The Classic route starts at the Scarth Wood Moor triangulation pillar (SE459997) and finishes at the bar of the Raven Hall Hotel (NZ980018). However, for all practical purposes the start and finish are the two Lyke Wake Stones: one at Sheepwash Reservoir (SE467993) the other at Beacon Howe (NZ970012).

The original intention was to allow people to choose their own routes, but sticking to the tops as far as possible, giving a broad band across the moors. To cross the B1257 Stokesley-Helmsley road between Clay Bank Top (NZ573035) and Orterley Lane (SE563985); the A169 Whitby-Pickering road between Sil Howe (NZ855027) and Saltersgate (SE852943); and the A171 Whitby-Scarborough road between Evan Howe (NZ923015) and the Falcon Inn (SE971981). Over the years a more definitive route has developed with only one or two alternative sections being used, but you are welcome to choose your own route if you want to keep off the well worn main one! Cycle crossings are not accepted, and are currently illegal.

Before attempting the walk please send a stamped addressed envelope for the latest club circular, which gives current information on the route. These are available from; **The Lyke Wake Club, PO Box 24, Northallerton, DL6 3HZ.**

All who do the walk and wish to claim membership of the Club must send a report of their walk, pungent or poetic, giving details of their route and times.

Reports have come in many formats, from a few lines to vast publications and photo journals. Interesting ones are quoted at Wakes, a few extracts from the late Ben Hingston's reports follow:

26 October 1974

For my 100th crossing I was blessed with a perfect day. It was mild and dry, very clear, and I had the wind behind me. There was hardly any litter, and very few people about. For the umpteenth time I rejoiced that Rosedale Chimney is no more. The wide horizon is vastly enhanced by its absence.

Left Osmotherley village at 5.00. Dew on the grass, but heather and bracken were dry. Used torch as far as the Shooting House 6.10, then followed the path. I carried my jacket all the way from Bilsdale Hall – it was too warm to wear it. Passed the lime heap 9.45. Misty drizzle for half an hour only. Stopped for a break at the Causeway 11.10. Detoured north of the bog, and crossed Rosedale Moor through short heather keeping well to the left of the muddy path. Blue Man 12.45, Stepping Stones 13.45, Lilla Howe 15.15. I stopped for a second break at Burn Howe, crossed Jugger Howe Ravine at 16.00 and came over Stony Marl Moor and the coast path to catch the bus from Ravenscar at 17.00.

12 April 1977

Dry west wind, after a cold damp Easter. Ossy 4.30, Dawn 5.20. Atmosphere very clear. Chop Yat 7.00, Blue Man 12.00, Lilla Howe 15.00. Fox moth caterpillars abundant. Finished 17.00 12+ hours.

8 June 1977

Mild west wind. Showers of drizzle and hail, but clear light and much sun. Ossy 5.10, Cock Howe 7.15, Botton Head 8.30, Blue Man 12.25, Eller Beck 14.25. Bus from Windmill 17.10 12 hours.

9 July 1977

L.W. Race. Cold damp fog all day. 9hrs 42 minutes.

28 August 1977

A wonderful day, and a long way round. No flies, no dust, no crowds, no litter and no mud. But there was heavy dew when I left Ossy in brilliant moonlight at 3.20. Chop Yat 6.00, Old Margery 9.00. Rest at Botton Cross 9.30 and came by Wintergill and Thackside to Birch Hall 13.00 to 13.30. Sil Howe 14.20. Slept in the heather 14.45 to 15.15. Came by York Cross to Maybeck because Ron lllingworth had left two cans of beer in a culvert there, to help him finish the Crosses Walk. But he never got that far. So I had them.

Then went south up the forest track onto Pike Hill, and came round south-east over Blea Hill Rigg and across through short heather above Hollin Gill to join the L.W. Path near Jugger Howe Ravine 17.10. Tea at Pollard cafe 16.30. There was no Sunday bus. But other dirgers had phoned Scarborough for a taxi to get them back to the start where their cars were parked. So I joined them in Raven Hall bar until the taxi came. They gave me a free lift to Ossy, and I cycled home by moonlight. 15 hours.

14 January 1983

Left Osmotherley 21.35. Slept in Swainby Shooting House for an hour. Moor filthy with pools of melt water. Cloud high, clear, constant visual delights. At Ravenscar a passing car whisked me into Scarborough to catch the 5 o'clock train and I was home by 19.30. Slept soundly after my 200th Crossing.

It can be seen from these reports that people still find interesting alternative routes, but since these were written the Ravenscar public transport has become of almost no use to Lyke Wake Walkers!

The Club makes a small charge for the black edged Cards of Condolence which successful walkers receive. This represents the cost, not of the card alone, but of the administrative expense of handling reports and keeping statistics. Please report all crossings to enable the Club to keep accurate statistics.

RACES AND RECORDS

In 1957 Arthur Puckrin did a west-east crossing in 10 hours 10 minutes and the double in just less than twenty four hours. By 1962 he had completed a triple crossing in 32 hours 15 minutes. Louis Kulscar and John Gray did four consecutive crossings in 78 hours in 1968, and so it went on.

In June 1965 consternation and controversy were aroused when Fred Day, a runner from Richmond, claimed a time of 4 hours 33 minutes. Athletic circles refused to believe that some mistake had not been made in the timing!

Generally the Club is opposed to making the walk into a race but in 1963 the organisers of the revived Osmotherley Summer Games, a village sports day of ancient lineage, asked the Club to organise a race over the route from Ravenscar. Arthur Puckrin won that first race in 1964 in 6 hours 30 minutes.

Ever since then the nearest Saturday to July 10th has become annual race day – the only day the Club will accept official record-breaking attempts, with Club marshals to authenticate times. The race is now held in the conventional direction, Sheepwash to the Raven Hall Hotel.

There are several classes for true walkers as well as runners. The current official records are as follows:

Male Winners

1964	A.Puckrin	6-30
1965	A.Puckrin	6-30
1966	G.Hall	6-30
1967	A.Puckrin	6-45
1968	P.Puckrin	5-16
1969	P.Puckrin	5-49
1970	A.Puckrin	6-59
1971	P.Puckrin	5-00
1972	P.Puckrin	5-27
1973	P.Puckrin	5-34
1974	M.Newell	5-22

1975	J.Williams	6-10			
1976	T.Florry	5-22			
1977	B.Dale	5-10			
1978	K.Robinson	5-21			
1979	J.Naylor	4-53	*Female Winners*		
1980	M.Garratt	5-09	1980	L.Lord	7-04
1981	M.Garrett	4-51	1981	L.Lord	6-38
1982	M.Rigby	5-09	1982	L.Lord	6-10
1983	G.Orchard	5-21	1983	B.Yule	5-57
1984	M.Rigby	4-57	1984	A.Collinson	7-43
1985	R.Mitchell	4-52	1985	L.Lord	5-56
1986	R.Mitchell	4-47	1986	E.Savage	6-52
1987	J.Rogers	5-24	1987	E.Savage	6-55
1988	R.Firth	5-15	1988	E.Savage	6-47
1989	M.Rigby	4-41	1989	C.Proctor	6-19
1990	M.Hoon		1990	E.Adams	6-32
	S.O'Callaghan				
	H.Bate	5-31			
1991	S.O'Callaghan		1991	C.Proctor	5-43
	J.Coulson	5-04			
1992	A.Carter	5-18	1992	H.Diamantides	5-30
1993	A.Jones	5-30	1993	S.Gayter	6-33
1994	B.Roberts	5-12	1994	J.Elliott	6-32
1995	A.Ward	5-34	1995	J.Nicholson	6-47
1996	A.Jenkinson	6-13	1996	G.Havis	7-42
1997	J.Rogers		1997	G.Havis	7-35
	A.Ward	5-30			
1998	C.Pattison	6-28	1998	G.Havis	8-05
1999	C.Pattison	6-39	1999	K.White	7-27
2000	E.Grant	6-17	2000	S.Burton	7-44

Many reports come in of youngsters having done the walk – the youngest, as far as we know, is a month short of seven years old – and from there up into their early eighties. Stan Garbutt of Stokesley was still entering the race until his late seventies – he died in 1999 otherwise he may have still been running.

Of the remaining elderly founder members, several undertook an official 40th anniversary crossing in October 1995, being joined along the way by other walkers culminating at an anniversary buffet at the Raven Hall Hotel.

Apart from the few people who have gone on into triple figures, very many men and women have completed the walk thirty, forty or fifty times.

Many disabled people have completed the walk, some in wheelchairs with assistance, some with artificial limbs, and one notable crossing by John Hawkridge in four days. A customer in the Queen Catherine thought that his slow progress on two sticks was the result of doing the walk. "Oh, no," said the barmaid, "he didn't walk any better than that when he started."

THE ROUTE

Many things have changed since the first edition of the official Lyke Wake Walk guidebook was published by *The Dalesman* in 1959. National Parks have become more powerful, people are more interested in ecological and environmental matters, permissive paths have evolved, village pubs have become fancy restaurants, estates have changed ownership and major changes are taking place in countryside access legislation.

Although Ravenscar and Osmotherley are still 67km apart, and not getting any closer, the walk has become very much easier: paths have developed across moors which were knee deep heather; the western section is paved and waymarked for many miles as part of the Cleveland Way; and generally route finding is not a problem.

The western section is not only used by the Lyke Wake Walk; several others including the Coast to Coast and the Cleveland Way follow the northern escarpment of the moors. It is shown on current editions of Ordnance Survey maps as the Cleveland Way, as far as Bloworth Crossing – just over a quarter of the way.

As this is a 'challenge walk', part of that challenge is to find your way from one end to the other. This brief route guide outlines what has become the *usual* route. However, at the time of updating this book, the club was amazed to receive a letter from an East Anglian man of the cloth, who claimed that the path from Osmotherley car park "…soon petered out. They were only able to cross the moor using a map and compass." Unable to find a path used by many thousands of people per year, a challenge they ultimately failed!

The walk may be done either way, of course, but with the prevailing winds being south-westerly a west-east crossing is usually easier. This guide takes you towards the North Sea from the Lyke Wake Stone at Sheepwash car park.

Although most people do not go to the triangulation pillar on Scarth Wood Moor, it is well worth visiting in clear weather conditions for long distance views towards the Pennine Hills: Pen Hill, Great Whernside, Buckden Pike, Mickle Fell and, almost 80km away in Cumbria, Cross Fell.

The path up to this viewpoint is clearly defined going west from the Lyke Wake Stone, uphill onto Scarth Wood Moor. From there the route goes north following the Cleveland Way back down to the Swainby road at Scarth Nick Cattle Grid (NZ473004).

To avoid this detour, follow the Swainby road from the Lyke Wake Stone for fifteen hundred metres to the Scarth Nick Cattle Grid. Turn right off the road, through a gate, and follow the wide forestry track eastwards around Coalmire to the Bill Cowley Memorial Stone at the top of the old ironstone miners incline.

This steep pathway takes you down into Scugdale. Cross the main access track leading from Swainby and follow the bridle path through woodland until open fields are reached on your left. Two clearly signed paths cross this field, taking you over a ford to join a surfaced road at Hollin Hill Farm. Continue on this to the main Scugdale road at Huthwaite telephone box (NZ493007).

Almost immediately opposite the road junction a gate leads to a footpath taking you towards the moor. On your right in woodland are the remains of the Ailesbury ironstone mines. Follow this path, bearing left along the lower edge of the forestry until a well-defined path goes up the nose of Live Moor. Some of this is a stone surfaced stairway onto the open moor!

Most of the path along this edge of the northern escarpment of the moors has been paved. The National Park Authority carried out this work between 1992 and 1999, and some of the earlier sections are now supporting vegetation, and are a good walking surface – and in poor conditions a good waymarker.

To your left above the village of Swainby you will see the remains of Whorlton Castle, one of the residences used by Edward II when on a hunting foray on these moors. Little is known of the route taken by his party but the King and his Court also stayed at Danby and Pickering Castles.

Many walkers use this ridge, a watershed between the Tees and the Humber; it needs little description; continue eastwards across Live Moor with its ancient field systems and onto the higher Carlton Moor, the home of an almost unused gliding club. The gliding club's wholesale levelling of the moor top in the early sixties is much regretted, although plant life is returning to the moor. Keep the club buildings to your right and the paved causeway takes you down to the road at Carlton Bank.

On your left, as you descend, are the remains of 19th century alum mining workings. The whole area was further environmentally damaged by motorcycle scrambling in the sixties and seventies. In 1997 the National Park Authority started a £1.8m refurbishment of the area, stabilised the eroding areas, carried out drainage and laid a geotextile surface to hold vegetation. The prognosis now looks favourable with vegetation slowly covering the erosion scars.

Once at the road (NZ523030) cross over the stile in the roadside fence and continue towards Cringle Moor. After a hundred metres, on your right, is one of the few new developments on the whole Lyke Wake Walk route: Lord

Stones Café, no doubt erroneously named after the old 'Three Lords' Stone', marking the junction of parish and estate boundaries – the three lords being Duncombe of Helmsley, Marwood of Busby and Ailesbury of Scugdale.

Continuing towards Cringle moor the path forks at the wall corner. A path on your left takes you around the front of the moor on the old jet miners' track. Or continue uphill for the true 'classic' route. Another fine viewpoint from the Alan Falconer memorial dialstone – take it with a pinch of salt, about seeing Ingleborough! The path continues over the summit of this, the second highest hill on the North York Moors, and drops down into the dip before Cold Moor.

Unless you are dedicated there is little sense going over Cold Moor. Continue on the old miners' track to Hasty Bank and the Wainstones – a popular climbing area. Again the choice is yours, over the top, or contour around the front, both paths are well defined. The one over the top is worth the little extra effort, and both lead you to the 'Robbie' memorial seat and down the stone steps to the main Helmsley road at Clay Bank Top (NZ573034), otherwise known as Hagg's Gate.

A little short of a quarter of the way to Ravenscar, but now you leave behind the wonderful long distance northerly views across Teesside and the Cleveland Plain, as you enter into the true 'Blackamore'.

Cross this busy road (B1257) with caution, and through the gate into

pasture (not the gate into forestry); the track you are on from Hagg's Gate is an ancient packhorse track. This footpath continues up the right-hand side of the wall, the old boundary of the medieval Greenhow deer park, until you reach the rocky outcrop. This, too, has had new pathways laid. Avoiding the scramble through the rocks, you will however miss the 'Sleepy Rosebud of Kermansha' memorial plaque. The stiff climb now levels out as it reaches the gate on the nose of Carr Ridge.

A little further on, to your right, is a long line of entrenchments along the western edge of Urra Moor; attributed locally to Cromwell they are at least medieval.

The well-used footpath continues, following the line of boundary stones as it climbs towards the Urra Moor firebreak which was bulldozed in the winter of 1960-1 to the dismay of Lyke Wake Walkers. Forty years on, it has become less of a scar on the landscape, and a useful path south-easterly towards the summit.

The Urra (O.E. *horh*, filth) Moor summit, Round Hill, the highest on the North York Moors, stands at 454m; on the left-hand side of the firebreak, close to this highest point, is the Hand Stone. *This is the way to Stoxla* and *This is the way to Kirbie* – a guidepost to travellers probably dating from 1711 when the Justices sitting at Northallerton on October 2nd "order that Guide Posts should be erected throughout the North Riding, all Surveyors in every Parish to order posts to be erected at the crossways according to the form of the statute."

A short way on, a much older stone, the Face Stone, with its sinister Celtic-style face deeply cut on the east side, mentioned in a 1642 Perambulation of the Helmsley Estate; "*Cookinge Rigg being the land of the Lord Duke of Buckingham on the East. And so goeinge N.wards vpp Barney Gill to the Streete Way. Then turning N.W. to the bounder called Faceston.*"

Both of these stones have been damaged in recent years.

A very ancient paved causeway runs through the heather on this moor top, some of it badly damaged when the firebreak was constructed. A Lyke Wake Club working party in the autumn of 1961 carried out some repairs and excavated considerable lengths previously unknown. It is locally known as the smugglers' trod, though no doubt of 17th century lineage, and maybe much older indeed.

After a further couple of kilometres eastwards on the firebreak the old Rosedale Ironstone Railway is reached, completed in 1861 to take ironstone to the then developing Middlesbrough. Turn right and follow the trackbed to the level crossing at Bloworth (NZ615015), where the unsurfaced Rudland Rigg road crosses the railway.

In the early days of the Lyke Wake Walk the old crossing gates remained, not the current modern one; and more importantly the continuing trackbed, which had its track removed in 1928-9, was not a public right of way. However, it is now a footpath, and the section around the head of Rosedale is currently being designated as a permissive path.

About five kilometres of monotonous walking towards Blakey Rigg, even worse in dark or low cloud when every one of the six or seven bends looks the same. A cinder surface either on high windswept embankments or in deep cuttings, lashing rain or knee-deep snow – horrible!

The latter part of this section has been controversial. In the early days walkers dropped down in to Esklets at the head of Westerdale, and up to Ralph's Cross; later a good track developed across South Flat Howe (O.N. *haugr*, heap) to the Old Margery Stone.

The Club tries to keep walkers on the railway track, until the Lion Inn, to appease the moorland estate management. Leave the railway track at the rear of the Lion Inn (SE675997) and take a short path following the intake walls to the pub. The estate management has proposed a further scheme which will require walkers to follow the railway track around the head of Rosedale.

From the Lion Inn the busy and fast Castleton road can be followed north to Ralph's Cross. There are about forty named crosses and guideposts on these moors, most of them very old indeed.

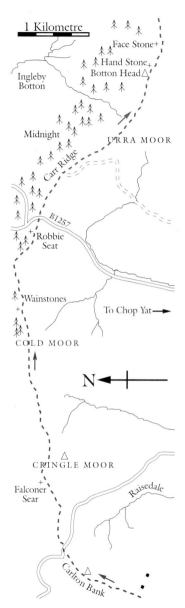

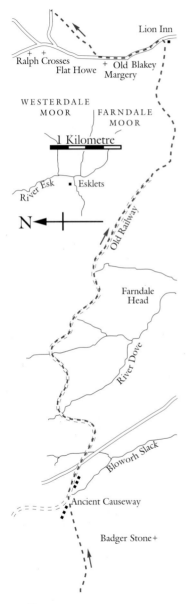

The symbolic badge of the North York Moors National Park depicts Ralph's Cross. From there follow the Rosedale road east to a point close to Loose Howe (NZ698011). Or you can drop down from the Lion Inn into the head of Rosedale and follow the railway around to the disused quarries below Loose Howe, and then climb up – an abysmal route, to be avoided!

Competitors in the annual Lyke Wake Race have tried this route. The route up to Loose Howe, though obvious when looking from the Lion, is quite unclear and in time is considerably longer.

The most pleasant way around the head of Rosedale is to follow the Castleton road from the Lion Inn to the boundary stones at Old Margery (NZ675013) close by; a footpath on the right-hand side of the road goes north-east along the white boundary stones to Fat Betty, otherwise known as White Cross (NZ682020), and continues south-westerly on the boundary to the road below Loose Howe.

Many of the burial tumuli on these high points have been unsystematically excavated in the 19th century, by 'pot-hunting' squires eager to increase their collections of cinerary urns. Loose Howe contained a massive boat-shaped coffin with inhumation and a bronze dagger.

Just before you reach Loose Howe where the Fryupdale road joins the Rosedale road, you will see a more modern megalith carved with a cross and the date ADMM: The

Millennium Stone, erected by the North York Moors Association in February 2000, no doubt taking a leaf out of the book of the Bronze Age people whose standing stones have acted as waymarkers for 3,000 years.

The inscription was carved by one of the Weatherill family, stone-masons in Eskdale since the late 1700s; much work has been carried out by this family over the years on churches, houses, memorials etc.

From Loose Howe the same white boundary stones are followed all the way to the next road at Hamer. Just past Loose Howe you cross the George Gap Causeway at the Cause-way Stone 1864, another of the old smugglers' and traders' routes. A lot of these boundary stones, including the Causeway Stone, have fallen over due to erosion in recent years, but the Causeway Stone has been re-erected a little way from the actual paved causeway.

From here, in the hollow ahead of you, lies a long section of deep bog and marsh. Over recent years it has been much drier – only ankle deep, not thigh deep. Is this better moorland management and drainage or global warming?

This whole section is a long stretch of bog and badly-eroded paths, caused by overuse in the sixties and early seventies. As early as December 1964, Bill Cowley issued a press release, part of which said; "The walk is a victim of its own popularity, there are times when I wish I had not started it, or could find a way to finish it."

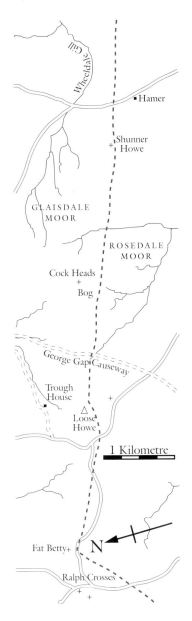

Fortunately the walk continued, and Bill even in later years accepted that much damage had been caused by overuse, but felt that it had been worthwhile for the pleasure it gave to almost 200,000 people.

Continuing eastwards, leaving this morass, you reach the halfway point. Many years ago a Lyke Wake wag raised a cairn and cross, with the following inscription:

> *Poor chap, he did try hard;*
> *He died to get his Dirger's card;*
> *Here he lies, now just half way;*
> *And here he'll stay for many a day.*

The cairn, like him, has long since sunk into the bog.

There were very serious moorland fires on Rosedale and Wheeldale moors in 1976. Expert prediction was that the moor would take ten years to recover; twenty-five years later the moorland vegetation and ecology is still far from recovered. Parts are now involved in experimental management schemes, the reason for small fenced-off plots.

The well-worn path continues to the large tumulus on Shunner Howe. To the north of here there are vast peat deposits controlled by the Danby Court Leet. It was in this area, whilst cutting peat, that the idea of the Lyke Wake Walk occurred to the late Bill Cowley.

From Shunner Howe the original route went to the ruin of the old inn at Hamer House; a flourishing inn until about 1870, when this minor road was one of the main roads south from Whitby. It was last occupied in 1936 and is now little more than a grass-covered pile of rubble at the roadside.

The path has since developed a little northwards to cross the road at SE744995 with suitable parking for support parties. You have covered 37km with about 30km to go!

Ahead of you lies the crux of the walk. You can see as far as RAF Fylingdales with its new early-warning pyramid – not as aesthetic as the old 'golf balls' which were derided by conservationists when they were built. Strangely, some people wanted them saved as listed buildings when the military made them redundant several years ago.

It's Wheeldale Moor ahead which can be tiring. Most of it is still a lunar landscape after the 1976 fires, a surface of rock and grit, complete with the remains of a burnt-out fire engine a little to the south of the route.

The white boundary stones are followed from the road, crossing several man-made walkways and sections with experimental path surfaces until a large marker stone is reached "Blue-Man-i'-th'-Moss" – but then the boundary turns almost south and you need to continue eastwards, into the unknown.

In all probability 'Blue Man' is from the ancient British; *plu*, parish, and *maen*, stone, a parish boundary stone. North of you lies the large Wheeldale plantation, and south the vast Cropton Forest. These, like a lot of the forestry on the earlier sections of the walk, were planted about forty years ago. Now, almost fully grown, and as a crop, ready to harvest – many areas are currently being felled.

The path from "Blue-Man-i'-th'-Moss" is a little vague, mainly due to lack of vegetation; the path generally contours around the south side of Wheeldale Gill, slowly dropping towards Raven Stones. It is this section of about three kilometres, taking you to the minor road west of Wheeldale Lodge at SE804983, that is the main navigational problem in poor visibility.

Cross the stile in the roadside fence and follow the path downhill through rough pasture to the Roman Road, 'Wade's Causeway' or the 'Old Wife's Trod' – a first century link from the Roman camp at Cawthorn and the coast at Whitby. Continue down the steep bracken-covered hillside to the stepping stones at Wheeldale Lodge. But a word of warning: Wheeldale Beck has been known to be impassable in times of flood! It is the main drainage for a large area of moorland and carries a considerable flow in flood conditions – if in doubt, do not cross.

Coming down from Wheeldale Moor you will already have picked your line almost straight up Howl Moor. Leaving the stepping stones the path starts a steep climb up onto

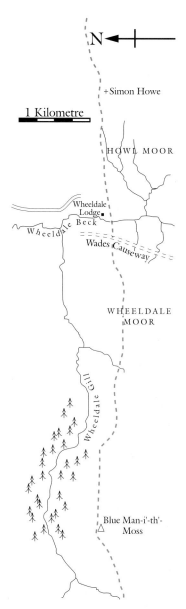

Howl Moor, which soon eases out to become a long steady drag up onto Simon Howe. In recent years this path has become very badly eroded and in wet weather almost a flowing stream – complete with very sticky yellow mud.

The path keeps south of the triangulation pillar on the summit and leads you straight to the tumulus on Simon Howe; the way the tumulus has been eroded or excavated in the past clearly shows the construction.

Some of the moorland tumuli were turf mounds with large stones acting as a kerb, others were mainly stone. Simon Howe shows these large orthostat kerbs. These burial and cremation sites are the most permanent remains of a nomadic people, the only permanent home they know.

Descending from Simon Howe the path is well defined, eroded and muddy down to the railway line at Fen Bog. The Whitby and Pickering Railway, engineered by George Stephenson, was opened on 26th May 1836 and closed 6th March 1965. Fortunately the track was never removed and due to voluntary effort and a charitable trust was reopened 1st May 1973. It is now operated on a commercial basis by the North Yorkshire Moors Railway. So don't be surprised to see *Sir Nigel Gresley* or *Vera Lynn* or a more mundane diesel-hauled train travelling along Newtondale, a glacial gorge.

Cross the track and follow the footpath through Fen Bog, a Yorkshire Wildlife Trust nature reserve. This footpath is one of the few places that it is easy to cross this important peat bog habitat.

A short walk on a hard surfaced access track and you join the Whitby-Pickering road (SE856982), at Ellerbeck Bridge, with three quarters of the route behind you.

For many years the section ahead of you, Fylingdales Moor, was in military occupation as a firing range, and later became the home of RAF Fylingdales with the 'cold war' Ballistic Missile Early Warning Station.

During the early period of the Lyke Wake Walk, aspirant walkers had to obtain indemnity certificates from the Ministry of Defence – old shells and mortars are still found at times. Later, permission had to be obtained to cross the moor from the Commanding Officer at the Early Warning station. These rules and regulations seem to have disappeared into the mists of time.

Turn left and walk downhill almost to the bridge on the main road, turn right following the MOD access track eastwards. The hard surface is soon left behind, but an obvious path on the south side of Little Eller Beck takes you towards Lilla Howe – visible on the skyline. Do not follow Eller Beck towards the Early Warning Station. Several small streams have to be crossed, and in wet conditions you can expect wet feet.

The footpath eventually joins the MOD perimeter track about three hundred metres before Lilla Howe. In poor visibility, take care not to turn right on the perimeter track at the fence corner; go straight ahead over a short section of heather to Lilla Cross. This cross is reputed to be the oldest Christian monument on the North York Moors.

During the period when Fylingdales was a firing range, the cross was re-sited a little way north at Sil Howe for its own safety; it was returned to its correct home in the summer of 1962 by the Whitby Council Surveyor, Graham Leach – himself a Dirger.

From here you can see the radio mast at Ravenscar, and it never gets any closer! Lilla Howe and Louven Howe a little way north have many access tracks and bridle paths radiating from them, including 'Whitby Old Road' and 'Robin Hood's Bay Road' – mediaeval roads. It's very easy to get onto the incorrect path in poor weather conditions. Head north-east from Lilla Cross and join the moorland access track, following this in the same direction to a point east of Burn Howe Duck Pond. Again, a further intersection of tracks; be careful, stay on the access track too long, and you will end up a long way south in the Derwent basin near to Brown Hill.

The correct one bears slightly left, two hundred metres past the duck pond, continuing about a kilometre to Burn Howe. Again, beware of taking a further access track leading you too far south-east towards Harwood Dale.

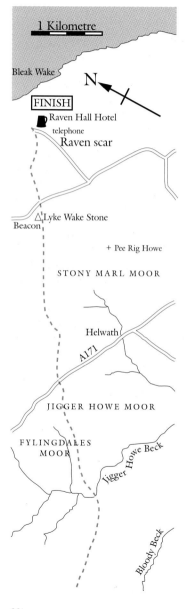

The path continues past Burn Howe to the top of Jugger Howe ravine, a very steep drop down on a deeply eroded gully. It is said by some that the agony of descending is only surpassed by the hell of climbing out of it.

It is these deeply-eroded sections which now produce the odd mortar bomb; it seems to have been a popular dumping ground for items found by bomb disposal teams, although few have been reported in recent years.

Jugger Howe Beck, an infant of the River Derwent, is even worse than Wheeldale Beck for being prone to flash floods. The whole of the valley bottom floods and even experienced Lyke Wake Walkers have difficulty. The late Ben Hingston, the Club's only Bi-Centenarian, claimed that he once swam across.

The stiff climb out does not help weary limbs, but a good hard-surfaced old military road soon has you at the lay-by off the Whitby-Scarborough road at Jugger Howe (NZ945003).

Weary walkers and fast traffic are a poor combination. The A171 is another fast and furious road; the entry to this lay-by is not in the best of places for visibility for either drivers or walkers. Once across the road a steep path goes up the side of the embankment onto Stony Marl Moor, and the Ravenscar radio mast still looks no closer!

To use Bill Cowley's words; "From there the final long mile of track goes straight across Stony Marl Moor to the Beacon" – there isn't much to add.

For those true stalwarts that want to continue to the traditional finish, at the Raven Hall Hotel, there are two alternative routes from the Lyke Wake Stone on Beacon Howe.

Either turn right on the road and follow it for almost a kilometre to the junction at the old windmill, then turn left following the road downhill into Ravenscar village. The hotel, with splendid views over Robin Hood's Bay, is situated at the end of the village on the cliff top.

A slightly shorter and more pleasant route is to take the footpath to the left-hand side of the radio mast, following an ancient sunken highway to the north-western end of School Lane, otherwise known as Robin Hood's Road, a bridle lane out of the village. Just before you reach the first house on this lane you will see the Ben Hingston Memorial seat; but no time to be sitting around now. Follow the lane to join the main road into the village.

The walk has always inspired those who complete it to peaks of literary achievement, like the party who "approached Simon Howe like Old Testament prophets arriving at Mount Tabor". The Duke of Devonshire claimed to be the oldest Duke to have done the walk. He was 60 and is rightly proud of his Lyke Wake Tie, as you will be.

An interesting photograph hangs in the Raven Hall Hotel showing the party assembled at Scarth Wood Moor before the very first crossing, and showing Tom Goulton presenting the Goulton Tankard on the steps of the Hotel after that historic event.

No doubt after the blisters have receded and the aches have worn off, you too will be writing wonderful reports and claiming membership of this somewhat strange club, membership gained only by ordeal.

THE SHEPHERD'S ROUND

The Shepherd's Round was devised by the late Bill Cowley in 1983.The original intention of the walk was to alleviate some of the overuse problems on other well-known walks on the North York Moors, especially the Lyke Wake Walk.

The walk is a sixty kilometre circular route, over ancient definitive footpaths and bridleways on the north-west corner of the North York Moors National Park with over 1500 metres of climbing. If it is your intention to use the Shepherd's Round as a 'challenge' walk the current time for walkers is about eighteen hours for completion.

Done at a more leisurely pace it is an excellent two-day walk with accommodation at several points.

This is a very hard walk, and it is suggested that large parties reconnoitre the route beforehand with a necessity for training and careful preparation.

Several parts of the route are not well defined and it is recommended to use the Outdoor Leisure 1:25000 map sheet 26 'North York Moors (west)'.

The Lyke Wake Club would be interested to receive reports of crossings and a three-colour woven badges is available.

The following details of the route will enable you to have an enjoyable but strenuous walk. As it's a circular walk there is no defined start or finish, but for the use of this guide Sheepwash car park has been chosen on the Swainby to Osmotherley road and we have completed the route in a clockwise direction.

> I dreamt last night of England and the rain,
>   grey clouds across the Yorkshire hills, and mist
> haunting the moors, curled low in every grain.
>   Close huddled sheep keeping bedraggled tryst
> behind a broken wall; a smell of wet heather
>   music of rushing streams, beat of the wind:
> One solitary shepherd "Mucky weather!"
>   "Aye, dampish, an' ah've three young lambs ti find."

Bill Cowley (while in India)

Well trodden and waymarked paths from Scarth Nick to Huthwaite, Live Moor and along the northern escarpment of the Cleveland Hills. The route of the Lyke Wake, Cleveland Way and Coast to Coast paths, much of which has recently been paved and surfaced. No problems should be encountered on this section to the B1257 Stokesley-Helmsley road at Clay Bank Top, the adjacent car park suitable for support parties. At this point you have completed the first quarter of the walk.

A shorter alternative, or 'Short Round', would be to turn off the Lyke Wake route at Carlton Bank and go south to Cock Howe and all the way down Bilsdale West Moor to High and Low Thwaites into Hawnby – a total of about 40km.

Continue eastwards on the popular route, climbing all the way to join the main access track. Passing the early 18th century 'Hand Stone' and the much earlier 'Face Stone' to the summit of Urra Moor, the highest point of the North York Moors, 460m, continue eastwards to Bloworth Crossing where the unsurfaced county road crosses the disused bed of the Rosedale railway. Turn south, following the unsurfaced road down Rudland Rigg. This is one of the many ancient highways traversing these moors, the old Thurchilesti road (old Danish).

After about two kilometres you reach the 'Celtic' Cammon Stone. This has had a later inscription added to it in Hebrew, probably, like other classical inscriptions on rocks in the area, by Strickland, a Bransdale man, who went to King's College, Cambridge and later became Vicar of Ingleby Greenhow.

A further kilometre past the Cammon Stone a footpath comes up on the left from Spout House in Farndale; to the right is the stump of Cockan Cross. Continue on the main rigg road to 635986 to join a bridleway to Bransdale (beyond which you can see 'Three Howes' on the skyline). This reaches the road just north of Cow Syke Farm. A choice here – either a field path via Bransdale Mill to Colt House (this should be avoided by large parties) or turn right and follow the road around the head of the dale past Cockayne. The road rises steeply to the moortop at 610965; ignore the signposted bridleway to Bilsdale, as this takes you back over via Tripsdale. Continue a short distance and a cart track comes up on your left from Low South House in Bransdale. Almost opposite this a less clearly marked footpath goes right, leading you west and then south-west across 'Botany Bay'.

Make sure you keep to the south of the boggy area, making straight for the intake wall corner, and follow the north side of the wall, down to the ruin of High House at the head of Bonfield Gill. There's a stream to cross which may be difficult in flood. Follow the footpath uphill keeping the intake walls on your left. After leaving the walls you cross an access track; the footpath becomes a little unclear but aim generally SW until you reach a line of shooting butts after about a kilometre.

From the northern end of these a good but narrow path goes west. Avoid dropping down the slope to your right. Presently you will see the intake wall and fence of High Crosslets. Head straight across the heather to the gate; until waymarked this path is difficult to find and to follow. An alternative is to keep on the Bransdale road for a kilometre to Bondfield Gill Farm. Here a bridle path is signposted, which starts off as a good access track to shooting butts. It continues as a bridle path to the High Crosslets gate.

From the intake gate the path goes down two or three fields aiming to the north of High Crosslets Farm, and comes out on a stoned road. This leads you around the farm buildings to the north and will shortly bring you onto the main Bilsdale road (B1257) just opposite the Fangdale Beck turning. Note the rare green telephone box – a few years ago it turned into one of those modern shower cubicles, but local opposition caused BT to replace this unusual feature of upper Bilsdale. A little over halfway.

The Sun Inn a little way south is a good stop for liquid (only) refreshments. Complete with an interesting 16th century thatched cruck building, the original Spout House Inn which closed in about 1914, now open to the public; and the headstone of 60-year-old huntsman Bobby Dawson in the front garden.

Bobby, who died in 1902, was the whipper-in of the Bilsdale Hunt which is claimed to be the oldest hunt in the country. He was buried at St Hilda's churchyard in Urra.

Keep to the south side of Fangdale Beck village, follow the road to Malkin Bower Farm. Just before the farm, a footpath on your right leaves the road at the second field fence; alternatively, continue to the farm and take the good track up to the old quarry.

Both of these paths reach to a moor gate above the quarry from where paths go right and left. The correct one leads to the right, which soon takes you west until you meet an access track running north-south along the Bilsdale West Moor ridge. Cross this and continue west until you reach the SE corner of the old Wether House intake walls.

It is reputed that Wether House was in fact an inn; William Ainsley of the Sun Inn says it was known as 'Widderiss' although Bill Cowley found no documents to support this when researching his 'Snilesworth' book in the early 1990s.

From here a somewhat obliterated path runs more or less down the Wetherhouse Beck to the intakes of the ruined Honey Hill.

Another ruin, Sike House, lies a little way north and beyond that the ruins of Low Thwaites and High Thwaites. These along with Wether House made this area well populated until as recently as the 1940s. Census returns for the latter part of the 19th century show that these isolated upland farmsteads supported many large families.

A somewhat better and firmer route may be found by turning left above the Malkin Bower quarry and following a line of shooting butts to the skyline near a prominent cairn. Then head straight for Honey Hill keeping to firm ground to the south of Wether House Beck and bogs.

From Honey Hill a good track is soon picked up which heads SW towards Moor Gate on the Osmotherley to Hawnby Road at 539917. From here you can either follow the road into Hawnby or turn right just after the cattle grid and follow the bridle path and footpath around the west side of Hawnby Hill to Hill End House and Hawnby, and its pub.

As the farm names imply, this area was very important to the shepherd-monks of Rievaulx Abbey, the first Cistercian house in the north. Three quarters of your walk is now completed.

From Hawnby take the road for Kepwick, passing New Hall. This road becomes an unsurfaced county road at Arden Hall.

This is a beautiful area of steep limestone valleys. There are many alternative paths but at this stage it is best to use the unsurfaced road which climbs steeply up to Kepwick Bank Top on the old Hambleton Drove Road. Possibly one of the oldest roads in England, a track since neolithic and Bronze Age times, and used more recently as a route south for the Scottish cattle drovers.

Turn north along the drove road, passing the remains of Limekiln House, an inn until about 1890. Turn left at the obvious track junction after Limekiln House and over Black Hambleton, once the haunt of an extraordinary but kindly witch, Abigail Craister.

Below Black Hambleton you join the Osmotherley to Hawnby Road and pass another old drovers inn, Chequers (refreshments but no ale). The road bears left to Osmotherley, the drove road continues north to Sheepwash, your starting point.

THE HAMBLETON HOBBLE

The Hambleton Hobble was devised by Paul Sherwood in 1983 as a means of removing pressure off other well-known walks on the North York Moors.

The walk is a fifty kilometre circular route centred around Black Hambleton, a 399m member of the Tabular Hills, built up on Middle Jurassic rocks on the western edge of the North York Moors National Park. These Hills offer very fine views westwards over the Vale of York to the distant Pennines and give the walker about 750m of climbing on this route.

For those wanting a shorter route it can be done as two thirty-kilometre circular routes.

The walk is unusual in that it links the village pubs in Osmotherley, Hawnby, Scawton and Nether Silton. Most of these inns have changed ownership in recent years, and will continue to do so – it's not known what some landlords think of walkers. In fairness to them please be considerate and don't litter public houses with rucksacks and filth from your boots!

The Hambleton Hobble was never intended as a speed challenge; currently walkers are taking about twelve hours to complete the route, some sections of it are not well defined or waymarked – although this is improving. It is recommended that you use the Outdoor Leisure 1:25000 map sheet 26 'North York Moors (west)'; the old 1" and 1:50000 maps are not suitable as parts of the walk are on arable land and these maps do not show walls, fences etc.

With the exception of a short section through Forestry Commission land the whole walk is on public rights of way. In normal conditions these forest tracks are open for public recreational use but please bear in mind that at times of serious fire risk they could be closed.

Several suitable points are available for support vehicles, all of which are within a short distance of either the B1257, A170 or A19; these points have been referred to in the text (*), and to preserve good relations with villages on the route none are close to habitation.

The Lyke Wake Club would be pleased to receive reports and comments on this walk and a woven cloth badge is available.

The following pages give details of the route; as it's a circular walk there is no defined start or finish, but for the use of this guide Osmotherley has been chosen as a start point and the route has been completed in a clockwise direction.

North of the Osmotherley war memorial you will see a Cleveland Way sign. This points you through a snicket that leads to Back Lane. Crossing Back Lane the footpath continues over fields and down to the footbridge over Cod Beck. Cross this and follow the path uphill to White House Farm; pass the farm buildings on your right to join the farm track to a green lane. Here the Cleveland Way turns right, you turn left; almost immediately turn right off the green lane onto a footpath to Greenhills Farm where you join the road.

From here follow the road uphill to the moor wall corner; an obvious footpath takes you to the old drovers' inn at Chequers.

Chequers, now a farm and café, is passed and you continue on the road towards Black Hambleton. A few hundred metres after crossing Jenny Brewster Gill you will see on your left a footpath sign; this points you south-east through half a kilometre of knee-deep heather to rejoin the road near to the remains of Robinson Cross(*). It is quicker and easier to follow the road to Robinson Cross – the choice is yours.

Two hundred metres past Robinson Cross a good track leads off, to the right, to the disused farm at Dale Head.

This section of the Hambleton Hobble from Dale Head to Hawnby is rich in history, many of the farm sites having been occupied since the Bronze Age.

From Dale Head walk down the field to your left to join the footpath which runs alongside the beck. Cross this on stepping stones and take either the footpath or the bridle path to Lower Locker Farm. If you use the lower of these two paths, the bridle path, you must turn right when you reach the wall before Lower Locker and follow the wall uphill to join the footpath.

If you have used the upper of these two paths, the footpath, it is less well defined, but aim for the gate on the skyline above Lower Locker.

The White Rose Walk crosses your route here, having descended from the north-east side of Black Hambleton.

Keep Lower Locker Farm buildings on your right and pick up the well-signed and used path to Far House and Cow Wath. At Cow Wath you have completed ten kilometres.

From the remains of Cow Wath an obvious path leads you through three small fields and into Greens Wood. The path continues through scrubby woodland to Brewster Hill. Keep the farm buildings on your left and drop down to Eskerdale Beck, crossed on a new footbridge. Go straight uphill keeping the electricity poles on your left and you will find a new stile to enter the woodland.

Turn left on the woodland path and follow it to Harker Gates and Mount Pleasant.

You pass through the farm keeping the main buildings on your left, and exit by the farm track. Note the 'Bull' warning signs. After a couple of hundred metres leave the farm track at a gate on your right and follow the field boundary to the gate behind Arden Hall. Use the estate track to Arden Hall, which in the mid-twelfth century was a Benedictine Nunnery. Nuns Well is in amongst the trees to your left, in a small hollow. Just before the Hall you will see a new seat erected in memory of some walkers killed in a road accident.

Turn left after the main agricultural buildings and pass Arden Hall, another of those buildings which claims to have had Mary Queen of Scots in residence. An impressive ancient yew hedge, and 'Kangaroos for the next 25km' sign, Join the road at the Hall gates(*).

Here the walk can be split into two circuits. If you are doing the shorter sections, turn right here and follow the unsurfaced road over the moor to Kepwick.

To continue on the full route, turn left, ignoring the woodland access track in front of you, and follow the road a short distance to the bridleway sign leading you to Coombe Hill.

On the gate entering Coombe Hill field you will see a sign indicating a permissive path (access agreement 006F3); this leads you around Coombe Hill, but the correct bridleway is more interesting.

The bridleway leaves the field on your upper left side adjacent to the site of Coombe Hill House, taking you around Coombe Hill and north of Carr Woods.

Joining the road at Church Bridge (*), it's now decision time: a short road walk to Hawnby Hotel or do you carry on? To carry on, after crossing Church Bridge take the immediate right fork into the churchyard of the 12th century All Saints Church.

Some interesting memorials inside, and an extract from the *Yorkshire Herald* of 23rd October 1916 – "Hawnby's Roll of Honour, A Proud Record of a Moorland Parish." Leave the churchyard by the small gate to join the road, and cross the River Rye on Dalicar footbridge.

Turn right after crossing the bridge. You will see the remains of an old shed in the field corner. Go uphill behind this and join the well-signed path to Sunnybank Farm.

The path continues, well signed, downhill to Daletown farm and then up to the road from Hawnby to Boltby(*). Vast panorama towards the north and east: if it's August a sea of purple when the heather is in flower; not far along the road towards Boltby is one of the North York Moors main climbing areas, Peak Scar, and some of the areas potholes known as windypits.

Turning your back on this view you have three kilometres of minor road walking, via Caydale Mill to Old Byland. At the T-junction in Old Byland ignore the small gate in front of you and follow the road to the right a hundred metres and take the left-hand gate into the steep-sided valley. You have completed twenty kilometres.

The path continues well defined to the top of Callister Wood and down to the stream in Nettledale. Cross the stream on the new footbridge and take the unsigned track uphill to the road at Scawton(*).

This walk started because two brothers had two pubs, one, the Three Tuns in Osmotherley, the other, the Hare Inn at Scawton. I was discussing with the Osmotherley brother the possibility of a new walk. He said, "Why don't you walk to our lad's place." The Hare Inn is at the far end of the village. Scawton church has an interesting plaque in the porch in memory of the crew of a Halifax bomber that crashed at Scawton in March 1945 on returning to RAF Elvington after a raid on Hagen. RAF Elvington is now an air museum near York, complete with rebuilt Halifax.You are now near enough halfway.

Behind the phone box in Scawton you will see a stile; take this, the start of a long section of arable field walking. Please adhere to the correct line of pathways, even more so when fields are in crop – and remember grass is a valuable crop. The path is well defined through Brignal Gill and Flassendale to Cold Kirkby, a village owned a hundred years ago by one of the Teesside 'Ironmasters', the Bolckow family.

Cross the village green and follow the bridleway sign outside East Farm, taking you up the farm drive to rejoin fields, to Back Lane. Almost straight across this lane and the bridleway contines over fields to join the road (*) near to Wethercote Farm. Turn left and follow the road to the T-junction with the old drovers road, which is surfaced at this point.

The Drovers' Road, or Hambleton Street, was used by Scottish cattle drovers for many years until the late 19th century, to drive cattle from Scotland to York and further south; many publications detail the history.

From this T-junction a path leads westwards over a couple of fields to the western escarpment of the moors. On a clear day you can see north west to Cross Fell in Cumbria, Mickle Fell at the top of Teesdale, west to Swaledale, Wensleydale, Coverdale and to Great Whernside between Wharfedale and Nidderdale and south-west to moors above Otley and Ilkley. Almost under your feet at this point are the remains of the Iron Age earthworks on Boltby Scar, but turn right and follow the escarpment to High Barn Farm where you descend on a well-signed zigzag path to Hesketh Hall. At this point you have done thirty kilometres.

Turn right and follow the road to the entrance to Boltby Forest (*). Stay on the main forest track, with the reservoir, originally constructed in 1885, on your right; when you come to the first obvious junction keep straight on, at the next obvious junction (489892) turn left, and after about ten metres turn right onto an unsurfaced forest track. It crosses a semi-surfaced track below Silver Nab and becomes a well-used and clear path. At Windy Gill the forest has been cleared and there are good views south to Eggborough and Drax power stations near Selby. You leave the forest at Gallows Hill and follow the bridleway down to Kepwick and further good views towards the Pennines. In early summer this area is awash with rhododendron blooms.

You join the road west of Kepwick (466907). It's at this point that you would restart the route, had you taken the shorter circuit from Arden Hall.

Well-signed and well-used path across fields to Nether Silton; just before the village an interesting stone marks the site of the old manor house. The Gold Cup public house is a few metres to your left.

In front you will see a small white gate. Go through this and follow another well-used path to Over Silton. You have now completed about forty kilometres.

A signed bridleway leads out of Over Silton and into forest. Confusion reigns: there are several routes through the forest. If you want to see the Hanging Stone (which you can hardly see for trees!) take the path that leads you around the western edge of the forest, otherwise use the main forest track to reach Thimbleby Moor and then turn left, downhill along the perimeter fence to join a path to Oakdale Woods.

Several new fences and signs in Oakdale Woods advising of 'no right of way' but they keep you on the correct path into Oakdale, descending between the two reservoirs, to the rebuilt Oakdale House – the keeper's house.

Take the access track out from Oakdale, crossing a bridge close to the lower reservoir and joining the road at Rose Cottage; turn left and follow the road to the bridge over Cod Beck, then turn right passing the old mill to rejoin your path back to Osmotherley below White House Farm.